GREAT WESTERN
PICTORIAL
No. 2

GREAT WESTERN
PICTORIAL
No. 2
THE HUBBACK COLLECTION

JOHN HODGE

DEDICATION

I obtained the Hubback collection through the services of my late friend Leighton David, a photographic chemist, of High Street, Barry. Gerald Hubback's widow had asked Leighton whether he knew of anyone who would like to have her late husband's railway photographs and negatives and he, knowing of my parallel interest, immediately asked me. I was delighted to receive a collection which contained so much material from before I was born.

Unfortunately, Leighton, like Gerald Hubback, died at an early age. It is therefore to the memory of Leighton David that I should like to dedicate this book. He did a great deal to help people interested in photography in Barry at that time. I also owe a great debt of gratitude to Gerald Hubback's widow for her foresight in passing the collection on through Leighton.

ACKNOWLEDGEMENTS

I would like to express my thanks to Mike Back and Richard Woodley for their help in tracking down photographic locations, and Ray Caston for providing detailed information on signalling, between Cardiff and Bridgend; to Peter Gray for identifying locations in the Taunton area and to Normal Carey, Bob Tuck and former Electricity Board Staff at Cardiff and Barry for help on biographical aspects. Most of the hard work on the captions has been undertaken by John Copsey, to whom I am extremely indebted.

Designed by Paul Karau
Printed by Amadeus Press Ltd., Huddersfield

Published by
WILD SWAN PUBLICATIONS LTD.
1-3 Hagbourne Road, Didcot, Oxon, OX11 8DP

Title page
Laira-based No. 6024 *King Edward I* pictured just to the west of Norton Fitzwarren with a down West of England express at 3.15 p.m. on Sunday, 10th September 1933, the last day of the summer timetable. The train is believed to have been the 10.50 a.m. Wolverhampton (via Stratford), with through coaches for Plymouth (front) and Paignton.

J.G. HUBBACK

The pre-'Castle' days of the Paddington & South Wales services, with No. 2934 *Butleigh Court* heading the 11.55 a.m. Paddington express past St Georges, just to the west of St Fagans, c. 1930. No. 2934 was transferred to Canton shed during July 1929, leaving for Swindon factory (then Old Oak shed) in November 1930, and was probably recorded here during that period. The 11.55 a.m. Paddington was a Milford Haven train, with the front portion (including the dining car) for Swansea, the middle section for Milford, and the rear for Pembroke Dock.

John Gerald Hubback (1903–58) was one of the fortunate breed of railway photographers who grew up with Great Western 'Saints' and 'Stars' in their prime, and saw 'Kings' and 'Castles' from new. He was in his prime from the mid-1930s until his early death at the age of 55. The overall quality of his work was consistently high from about 1934, and it is from that decade, to which this book is largely devoted, that his material now contains the greatest interest.

He took his first railway photograph on Tuesday, 21st May 1929, with a 1A size camera at Pontsarn Signal Box, between Cardiff and Llantrisant, of the double-headed 11.55 a.m. Paddington to West Wales express, using 1/50th sec. at f7.9, keeping the image sharp by releasing the shutter while the train was a good distance away from the camera, but creating interest at that distance by having it passing a bracket signal; he had good perspective from the start. It is probable that he was at Pontsarn in connection with his work on lineside equipment for the Electricity Company as he was back there again the following Saturday, this time gaining proof that 1/50sec. would not 'stop' a Saint doing 60 mph, and this probably led him to purchase a 116 size camera with a shutter speed that would do the job. He progressed rapidly from attempting to do the impossible with a slow shutter speed, quickly mastering train speed, and by 1934 he was producing high quality action pictures on 116 size negatives, changing to

16 on 120 from the mid decade, but retaining his larger format camera which he was to continue to use after the war.

By the mid-1930s, he had become one of the leading photographers on GWR territory, with work published from time to time in the *Railway Magazine, Railway Gazette* and later in other publications, but was never well known. He was always spurred by an urge towards technical perfection, his motto being 'Only the best will do!' He applied this philosophy both to the technical quality of his negatives, and to the excellence of his composition and backgrounds, most of his shots being taken in the countryside. This book is intended as a memorial to him as a very good (but little known) railway photographer whose work deserved during his lifetime, and still deserves today, to be far more widely known and acclaimed.

Gerald Hubback was an electrical engineer by trade and worked for the Electricity Company. He enjoyed the fortunate position of living in South Wales, at Barry, and working at Taunton from the late 1920s, where he lodged in a house overlooking the main line, from where he often saw the 'Torquay Pullman' during its short life. He travelled frequently at weekends between the two areas by using the P. & A. Campbell steamers across the Bristol Channel between Barry and Weston/Minehead, and his work during the 1930s is equally spread between the West of England and South Wales main lines, with a galaxy of

1

At times, the London expresses were double-headed west of Cardiff, though the extra engine was probably attached in order to position it for a further turn, or as a means of returning it to shed, in either instance to avoid light engine mileage. The leading engine, 'Saint' No. 2952 *Twineham Court* from Landore, was piloting an unidentified 'Hall' on the 11.55 a.m. Paddington to Milford Haven, during 1931. At this time, Canton worked the train from Paddington, with either a Canton or Landore engine onwards from Cardiff.

Another double-headed train, this time in the West Country, approaching Cogload. A 15-coach up train, probably for the North & West route, is seen behind a pair of 'Stars' – Nos. 4047 *Princess Louise* and 4062 *Malmesbury Abbey* – in the summer of 1931. No. 4047 was a Bath Road engine, whilst 4062 came from Shrewsbury, a shed from which engines worked daily through to Bristol, and regularly to Newton or Plymouth, especially during summer weekends. In the late 1930s, No. 4047 was allocated to Weymouth, and regularly worked the 11.45 a.m. Weymouth, returning with the 6.0 p.m. Paddington boat train. Evidence of construction work for the new flyover road may be seen in the foreground.

'King' pictures in the Cogload, Taunton, Norton Fitzwarren, Whiteball area, and a fascinating collection of 'Saint' and 'Star' pictures between Cardiff and Bridgend, during a period when he was virtually alone there as an active railway photographer. After the war, he resumed his photographic exploits from 1946 when he produced some of his best quality work, mostly in South Wales, where he now worked, often again on 116 size negatives.

It was probably through his work on electrical installations on railway property that he gained so much access to the lineside for photography, and was able to seek out the best vantage points, some possibly up telegraph poles. He had several favourite spots of pictorial excellence and would often take a succession of shots in the hope of producing the perfect picture, in so doing building up a fascinating collection not only of locomotive but also stock types. In the West, some of his best shots were in the area of Whiteball Tunnel, between Wellington and Burlescombe

Norton Fitzwarren was another favourite spot, with its conveniently located long-span footbridge, as also were Cole, Cogload and Creech. At these and other places he saw new 'Kings' and 'Castles' making their first appearances and photographed the legendary 1930's Summer Saturday succession of services from Paddington to Cornwall and Devon, often loaded to 15 bogies, the newest GWR coaches working alongside the older clerestory stock.

Between Cardiff and Bridgend, he spent hours at the lineside where the Barry Railway Viaduct crossed the main line just west of St. Fagans and found a host of good vantage points between Ely and Llanharan to record the galaxy of 'Stars' and 'Saints' which worked the London services in the early 1930s, before giving way to 'Castles' and retiring to secondary duties until the early 1950s. He witnessed the era when London services were still double-headed beyond Cardiff by 4–4–0s and 'Saints' or 'Stars', but this was at the dawn of his interest and few shots are usable.

Following his death in 1958, his widow wondered what she could do with his collection, and I became the happy recipient through a mutual friend. Quite a few negatives were missing from the collection I received (including many modern ones), and efforts to locate these have proved abortive. Also, many captions for the 1936–39 material were missing and much of the effort to produce this book has centred on trying to define the location and approximate date of much of the late 1930s material. Though I am confident that the location of all the material has been accurately established, the exact date of many shots has been impossible to fix, and only an approximation can be given, assisted on occasions by some aspect of the engine or train.

Some material is historically very valuable, such as a few pre-1934 pictures at Cardiff General and Canton before the rebuilding and re-modelling took place, photographic records of which are very rare. There are also prints of the new layout going in at Cogload in 1932. The photograph of 2900 *William Dean* passing Peterston with an up London service in 1931 when based at Landore is the only known picture of this engine in South Wales. The collection also contains shots of the converted 4–4–2 'Saints', taken in the early 1930s when many were South Wales-based plus a few 'Saints' and 'Stars' with a 4–4–0 pilot west of Cardiff, and some valuable records of coaching stock formations during the whole of the 1930s decade.

This volume concentrates largely on his work on the South Wales and West of England Main Lines during the 1930s and postwar years. I hope that it will provide an interesting and valuable photographic record of GWR main line activity in these areas during that period and will bring back happy memories to those people who, like Gerald Hubback, were fortunate enough to see those sights for themselves, and will show younger readers what they missed.

John Hodge,
Haywards Heath,
West Sussex

A late-1930s view of the east end of Bridgend station, with Canton 'Castle' No. 5046 *Earl Cawdor* passing through the station with the 8.25 a.m. Fishguard Harbour to Paddington (12.10 p.m. ex-Cardiff), normally comprising around ten through vehicles from Swansea (front, including a dining car), Pembroke Dock (centre) and Fishguard (rear). The train ran non-stop from Port Talbot to Cardiff, and was due through Bridgend at 11.41 a.m. In addition to the usual weekday vehicles, on summer Saturdays the train was scheduled to convey three or four through LMS coaches from Tenby to Liverpool as far as Carmarthen, and three additional GW vehicles from Pembroke and Tenby to Paddington. The mixed collection of coaching stock in the Vale of Glamorgan bay on the left, probably formed the next departure to Barry. These trains were generally worked by ex-Taff Vale Class 'A' 0-6-2Ts from Barry shed, which handled most of the passenger working between Bridgend, Barry and Cardiff at that time, supported by the occasional '56XX'.

BRIDGEND

The first down daytime Paddington service of the day, the 8.55 a.m. Paddington to Pembroke Dock, running past Bridgend East box behind Old Oak-based No. 5014 *Goodrich Castle*, photographed from the bridge at the east end of the station. The Vale of Glamorgan line ran into the station bay platform independently of the main line (immediately in front of the East box), but was connected to it in the vicinity. Beyond the cutting, the main line curved gently to the left, whilst the Barry line ran straight on, connecting with the Vale of Glamorgan line (via Coity) that avoided the station. The 8.55 a.m. Paddington, with through vehicles for Swansea (front), Pembroke Dock, and Neyland (rear), was due to pass through at around 12.15 p.m., running non-stop from Cardiff to Port Talbot, though the 11.55 a.m. and 3.55 p.m. Paddington expresses both called at Bridgend. In the late 1920s, an Old Oak engine was scheduled to work through to Carmarthen, but in the mid-1930s, records show that Landore shed was also involved with this train. The siding to the right was used by Wagon Repairs Ltd. (amongst others), a very large concern with depots all over the country – such repair firms were a common sight in South Wales, which handled a vast number of goods vehicles.

LLANHARAN

With the building of the 'Castles' and 'Kings' in the latter half of the 1920s (most of which were allocated to Paddington and West Country depots), Canton received six 'Stars' in the autumn of 1926 cascaded from the company's 'premier' route for use on London services. In 1927, Carmarthen also received a small allocation, followed by Landore in 1928, thus bringing the class back to South Wales in some numbers. No. 4003 *Lode Star* first arrived during April 1928 when it was allocated to Canton. The engine moved north a year later, but reappeared at Landore in July 1934, and remained there until withdrawn from service in July 1951 for preservation. It is seen at the eastern end of Llanharan one afternoon in the postwar era with an eastbound fast, which it probably worked as far as Cardiff, though there were occasional workings beyond, even to Birmingham on Saturdays.

Many train reporting numbers were reintroduced for the summer of 1946, continuing the prewar arrangement that had been largely (though not entirely) disbanded after 1940. Train No. 835 was the 12.0 noon Neyland to Paddington, the number being carried each weekday on the locomotive's smokebox onwards from Swansea, and is seen here at speed to the east of Llanharan, possibly behind Landore-based No. 4078 *Pembroke Castle*. This train was scheduled for five coaches from Swansea (including a dining car) at the head, six from Neyland, and three from Pembroke Dock at the rear, maintaining the character of the prewar services.

The London to West Wales boat service (Train No. 958), hauled by Landore-based No. 5013 *Abergavenny Castle*, approaching Llanharan c.1947. In the winter schedule for 1947, this train was identified as the 11.35 a.m. Paddington to Fishguard Harbour (Tuesdays, Thursdays and Saturdays only), which connected with the 9.0 p.m. sailing from Fishguard to Cork. Loaded to 14 vehicles on this occasion, the train was nominally scheduled for a four-coach section for Swansea at the head, and six-coach portion (including a dining car) for Fishguard. Previously, the 11.35 a.m. had been a Carmarthen train.

Another of the Landore 'Stars', No. 4048 *Princess Victoria*, looking rather the worse for wear as it passed under the occupation bridge and commenced the descent towards Cardiff at the head of an up stopping train, possibly the 10.12 a.m. Carmarthen, in early BR days. Along with Nos. 4003, 4023, 4039 and 4050, this engine was transferred to Landore in the mid-1930s and remained there until withdrawal, which, in the case of No. 4048, was January 1953. The well-patronised Rhondda company bus in its dark red livery was on the A473 Bridgend, Llantrisant and Pontypridd road, about three-quarters-of-a-mile to the east of Llanharan village, which was also situated on the north side of the line.

Hubback went on to record many 'Saints' and 'Stars' in the Peterston area, a bicycle ride away from his home in the Peterston area, a bicycle ride away from his home at Barry. 'Saint' 4–6–0 No. 2915 *Saint Bartholomew*, then a Canton engine, is seen on a westbound stopping service near Miskin in the latter 1930s; this engine was transferred to Cardiff in June 1934, and remained there until May 1939. On Monday, 9th July 1923, *Saint Bartholomew* hauled the inaugural 'Cheltenham Flyer' from Swindon to Paddington (77¼ miles) in 72 minutes with a load of nine coaches (250 tons), three minutes under schedule. With Engineman Hopkins and Fireman Bailey on the

MISKIN AND PONTSARN

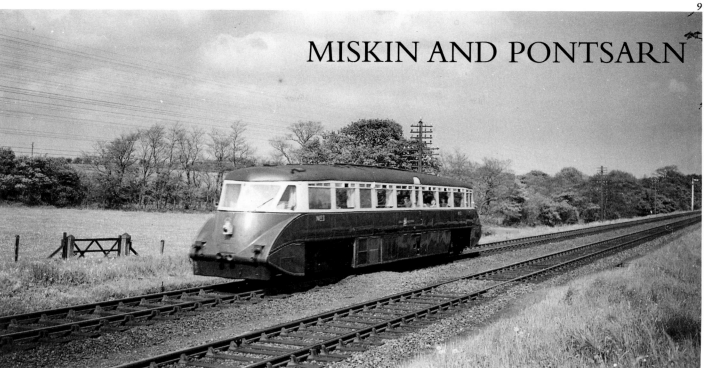

The success of the first AEC diesel railcar, introduced into traffic in December 1933, encouraged the Great Western to build a further three. These were different from the first in having two engines and incorporating a buffet counter, and were introduced to operate a pair of 'businessmen's' expresses between South Wales and Birmingham, a service which was inaugurated on 9th July 1934. From the late 1930s, the normal running condition appears to have been without panels over the bogies, which doubtless made maintenance easier. Car No. 3 is seen here near Miskin c. 1947 on an up service. Transferred to Landore in late 1946, No. 3 was used on the 9.35 a.m. Swansea to Cheltenham and 2.15 p.m. Cheltenham return, with an advertised buffet service, along with Nos. 2 and 4. Newport (Ebbw) and Landore cars also appeared on three other trains in each direction on this section of the line at this time.

Landore 'Star' No. 4023 *Danish Monarch* heading a down fast during the latter 1930s, comprising two passenger brake vans and three coaches. This engine was allocated to Old Oak from new in June 1909 until late 1923, working very largely on West Country turns, though with regular forays into South Wales; she was recorded on the 8.45 a.m. Paddington to Fishguard passenger with a load of 13 coaches on 31st July 1909. Canton received its first 'Star' in 1908, and by 1910 had four of the class, though the arrangement subsequently changed, with only Fishguard (Goodwick) having an allocation in the Principality by 1914. No. 4023 was subsequently transferred to Landore in January 1932 for a period of about 18 months, and again from December 1935 until withdrawal from traffic in July 1952. The use of white-painted engine headlamps and train tail lamps was authorised by Collett in December 1936, though the changeover from red was to be 'gradual'.

Another Canton Castle, No. 5080 *Defiant*, on a Paddington to West Wales express, passing through the beautifully wooded area near Miskin in 1946. The cutting affords a slightly elevated vantage point, whilst the slight bend shows up the formation to best advantage. This engine arrived at Canton shed from Old Oak Common in August 1940 as *Ogmore Castle*, receiving her distinctive 'Aircraft' series nameplate in January 1941; these commemorated twelve types in service with the Royal Air Force and Fleet Air Arm at the time. The first coach appears to have been in the wartime 'red-brown' livery, whilst the third vehicle looks like a rebuilt 'Dreadnought' restaurant car. The train was probably the 11.55 a.m. Paddington, with the first five vehicles destined for Swansea, the following five for Pembroke Dock (MWFO)

Canton-based No. 4901 *Adderley Hall* was well turned out for this down stopping service, probably the 1.8 p.m. Cardiff to Swansea. Hubback photographed trains in this area during the early stages of the war, and this may well be an example of his work at that time. In the early 1940s, this train was scheduled for a 4-coach set (Van Third, Compo, Third, Van Third) with an additional Third, and on this occasion was also conveying a Siphon 'G' and a Passenger Brake at the front. In 1943, the engine used had earlier worked the 8.0 a.m. Birmingham to Cardiff train (after an overnight stop at Tyseley), and would return to Cardiff that evening with the 5.50 p.m. Swansea. No. 4901, with the cab windows plated in, was paired with a 3,500-gallon tender, from new in 1928 until June 1947. *Adderley Hall* went to Cardiff from Truro in June 1941.

Much of the local passenger traffic between Swansea and Cardiff in postwar years was worked by 4–6–0s of the 'Star', 'Hall' and 'Grange' types, though with 'Castles' on a few. 'Grange' class No. 6857 *Tudor Grange* from Landore shed is shown heading down the bank with what is believed to have been the afternoon Swansea (ex-Carmarthen) to Gloucester and Cheltenham train. There were still a number of clerestory coaches in service during the postwar years, as witnessed here, whilst an inside-framed 'Siphon G' can be seen at the rear. No. 6857, again with its cab windows plated in, was also used on perishable traffic, and was recorded on the 1.15 a.m. Swindon to Whitland milk empties on 4th November 1947.

Another delightful woodland scene near Miskin with the 12.0 noon Neyland to Paddington service in 1946 behind a less-than-pristine 'Castle', No. 5015 *Kingswear Castle*. This engine was stationed at Stafford Road, a regular performer on North expresses out of Paddington at the time, but is seen here on a turn usually worked by a Landore engine. The 'Castles' were permitted to haul 455 tons (around 15 coaches)

The countryside between Llanharan and Ely was quite unspoilt, giving little indication of the heavy mining industry to be found only a short distance away. The gently sloping bank with the backdrop of trees to the west of Miskin Crossing offered a delightful setting for this picture of oil-burning '28XX' Class 2–8–0 No. 2864, passing through with a down Class 'H' freight during early summer. The wagons at the front may have been carrying scrap metal, a common 'ingredient' for the various metalworks in South Wales, whilst those immediately beyond were loaded with timber, possibly pitwood for collieries, with vehicles conveying general merchandise at the rear in this instance. The process of conversion of the '28XXs' for oil burning commenced in late 1945, with a total of twenty being converted in the period up to the summer of 1947. No. 2862 was converted in February 1946 and allocated to Llanelly, being renumbered 4802 that December. The engine was reconverted for coal-burning in September 1948.

4–6–0 No. 4094 *Dynevor Castle* was transferred to Canton shed in November 1937 after spells at Old Oak Common (1926–29), Newton Abbot and Laira (1929–37). As an Old Oak engine, she was primarily employed on West Country services, but was recorded on the 12.30 p.m. Paddington to Swansea on Sunday, 15th January 1928. Even when formally housed at Canton, 4094 occasionally wandered from her normal duties, for instance being utilised on the 4.5 p.m. Paddington to Birkenhead train (probably as far as Wolverhampton) on Saturday, 10th September 1938, no doubt replacing a failed Stafford Road engine whose turn it generally was. In this view, *Dynevor Castle* had just passed Miskin Crossing with the 8.55 a.m. Paddington to Pembroke Dock, c. 1946, which in postwar days was scheduled for four

A '43XX' class 2–6–0 hauling the 1.5 p.m. fast Swansea to Bristol train at the eastern end of the Miskin Loops, near Pontsarn box, on Sunday, 1st November 1931. This train, formed of elderly stock, was scheduled for a 'Lav. D' set at the time (Van Third, Compo, Compo, Van Third), though it had been strengthened with additional vehicles. The goods loops, opened to traffic in 1914, had been extended to a little over a mile in length during 1921, and were thereafter rated at '4 trains of 60 wagons, each with engine & van'. Although quadrupling had taken place between Cardiff and Newport in 1898, the remainder of the South Wales line through the industrial areas was double, with goods loops at strategic points.

Miskin Loops, with a view taken from Pontsarn signal box showing a seven-coach 2.5 p.m. Swansea to Gloucester service headed by a 'Bulldog'. At this time, there were around a dozen engines of this class at Canton and Landore, and they would doubtless have appeared regularly on the local passenger services. This was one of the photographs taken by Gerald Hubback on his first day out at the lineside with a camera, on Saturday, 25th May 1929. Taken at ⅕₀th sec. at f7.9, it shows an early attempt to 'stop' a train at speed with too slow a shutter speed, a problem he was to address very quickly.

An up freight, formed mainly of coal traffic, entering Peterston Loop, immediately to the east of the station in 1931. The ROD 2–8–0 was being assisted by Canton 'Bulldog' 4–4–0 No. 3441 *Blackbird*, which spent the period from 1927 until 1932 at that shed. In the 1920s, there were more than 100,000 mineral wagons working in the South Wales coal trade, largely owned by or hired to colliery proprietors, coal factors, merchants and other distributors. Peterston Loops were opened around 1896, and like those at nearby Miskin, were extended soon after the Great War (1922, in this instance). The signal box, Peterston West Jct., survived in operation until 1966.

PETERSTON LOOPS

The first of the 'Saint' class engines, No. 100 (later No. 2900, *Dean*, and *William Dean*) entered traffic in 1902, working largely out of Bristol and Wolverhampton until the latter part of the Great War. In October 1918, the engine was transferred to Canton and then to Pontypool Road, where it remained until late 1923. The engine returned again to those sheds from April 1925, before moving to Landore in February 1930. No. 2900 was photographed here on Friday, 7th March 1930, at Peterston Loop, working the 7.55 a.m. Pembroke Dock to Paddington train. This service was scheduled to convey vehicles from Swansea (front, including a dining car), Pembroke Dock, and Fishguard (rear), with a 'Siphon G' from Llanelly on the tail. No. 2900 was transferred to Chester in December 1931, being withdrawn from service six months later.

Landore 'Saint' No. 2908 *Lady of Quality* pictured at speed through Peterston Loops with the 8.55 a.m. Paddington to Pembroke Dock express, with through coaches for Swansea, Pembroke and Fishguard, on Saturday, 25th May 1929. The engine was passing over the junction with the ex-Barry Railway yard, through which coal traffic from the Bridgend and Llantrisant Valleys was conveyed to Barry Dock by way of Drope Jct. and Cadoxton. It is likely that the loaded Cambrian Colliery Co's. wagons were temporarily stabled on the loop, waiting to go across to the Barry yard, as they feature in other shots taken that day and do not seem to have been on a through train to Cardiff. The Cambrian company's offices were at Cardiff, whilst the collieries were located on the Ely Valley (Clydach Vale) line, the wagons probably joining the main line at Llantrisant.

By the late 1930s, 'Castles' were largely in charge of the main London expresses, but 'Saints' and 'Stars' were still very much in evidence on secondary services (and to a lesser extent, on expresses too). In this photograph, 2922 *Saint Gabriel* (Pontypool Road shed) was working the 10.55 a.m. Kensington to Whitland milk empties through St Georges at the point where the line skirts the country road. In the early 1930s, empty milk vehicles were for the most part destined for the depots at Carmarthen and Whitland, but later these were joined by others at Ffairfach, Pont Llanio and Felin Fach.

ST. GEORGES

This photograph, taken around 1929, shows 'Bulldog' 4–4–0 No. 3397 *Toronto* assisting 'Star' class 4–6–0 No. 4031 *Queen Mary* near Peterston, with a Paddington to West Wales express. No. 3397 was at Landore shed from 1928 until March 1930, and No. 4031 at Canton between April and December 1929, during which period this view was probably taken. The train may have been the 11.55 a.m. Paddington to Milford, with sections for Swansea (front), Milford and Pembroke Dock (rear).

No. 5038 *Morlais Castle* heading west through St Georges, past Peterston East down distant signal, with a down express, probably in the early war years. Judging by its formation, the train may have been the 11.55 a.m. Paddington to Neyland, due through the area at about 3.20 p.m., with the first five vehicles (including restaurant car) scheduled for Swansea, the next five for Neyland, and a Siphon 'G' for Carmarthen at the rear. The engine was from Old Oak, and surviving records show that the shed frequently used this engine on the Welsh runs. Engine rosters were seemingly changed rather more in wartime than in peace, and in the latter half of the war, the 11.55 Paddington duty was invariably worked by Landore 'Castles', with a variety of balancing turns. Entering traffic in June 1935, this engine spent her entire Great Western career and most of her BR days at Old Oak, finally moving to Shrewsbury in 1958.

Canton 'Saint' No. 2979 *Quentin Durward* is seen here working a down morning excursion near St Georges in the late 1930s, probably a Rugby special to Swansea, as one international each season was held there until 1954, and this photograph was one of a sequence of six. The number of specials run for sports events at this time could vary between one or two for an important local fixture to around forty for a Wembley cup final involving teams from Great Western territory. Rugby internationals similarly tended to generate heavy traffic in South Wales, whilst purely local events demanding additional passenger conveyance were often catered for by strengthening ordinary trains.

'28XX' class 2-8-0 No. 2863 was converted to oil burning at Swindon factory in April/May 1946, and was renumbered 4805 in that November; the '28XX' class oil-burners were renumbered into the '48XX' series in November and December 1946 (4800–06, 4850–52), or later upon conversion. The engine was allocated to Llanelly during 1946 (one of five) whilst a further five were at Severn Tunnel Jct. The engines were rostered initially in one sequence for the 8.55 p.m. Llanelly to Gloucester and the 8.45 p.m. Gloucester to Llandilo Jct (with connecting sectors for the Severn Tunnel Jct engines), and another for the 12.30 p.m. Llandilo Jct to Swindon, the 6.43 a.m. Swindon to Severn Tunnel Jct, and the 6.55 p.m. Severn Tunnel to Llandilo Jct. Here, No. 2863 is pictured passing St Georges in 1946 with a down Class 'H' mixed freight which included a number of fish vehicles, doubtless being returned empty to Neyland for Milford.

The use of 2–6–2 tanks on the main line was not very common in this part of South Wales, though they had worked through the Severn Tunnel as pilots (termed bankers) for many years. A member of the '41XX' series is seen here passing St Georges with a down local 4-coach set (partially marked 'Valleys'), plus a strengthening vehicle at the rear, probably in the early war years. The engine number appears to be in the '413X' series, which did not enter traffic until after the outbreak of war; No. 4138 was allocated briefly to Cardiff in January 1940, whilst other examples of that number series around that time were to be found at Severn Tunnel, Pontypool Road, Neath, Swansea and Llanelly. The train is believed to have been the 5.12 p.m. Cardiff to Porthcawl service.

Another view taken to the west of St Georges, probably in early postwar years. It shows Canton stalwart 'Saint' No. 2943 *Hampton Court* working a down stopping service, which could well have been the 9.55 a.m. Cardiff to Swansea (6.55 ex-Taunton). This train was always well loaded with the last of the overnight vans for Swansea and West Wales received into Cardiff. An interesting variety of passenger stock may be seen, representing four or five decades of Great Western coach design. No. 2943 moved to Canton in November 1941, and remained there until withdrawal in January 1951, though she had previously spent a two-year spell there in 1923–5. The 'shirtbutton' motif is still clearly visible on the 3,500-gallon tender.

The sweeping curve from St Georges, which passed under the Barry Railway viaduct on the western approach to St Fagans, was a favourite spot for Gerald Hubback. Standing at the west end of the viaduct bank, looking west, the little church of St Georges stood out on the south side of the line, whilst to the north (and just off the picture to the right) the former Barry railway line swept round from St Fagans to Tynycaeau Junction en route to Pontypridd. In the first view at this location, 'Saint' No. 2954 *Tockenham Court* (of Swindon shed) is shown with what may have been the 8.10 a.m. Carmarthen (9.45 Swansea) to Gloucester service probably during the early war years. The first two coaches, paper labelled on the windows, appear well filled, probably with a special party.

AROUND
ST. FAGANS

The point from where the photograph on the previous page was taken can be seen in this view of a down train coming round the curve on a westbound service, with the top of the Barry viaduct just visible in the background. Here a gleaming Tyseley 'Saint', No. 2930 *Saint Vincent* was easing round the curve with what could have been the 9.20 a.m. Birmingham to Carmarthen train in the late 1930s. During the latter 1920s, through engine working between Birmingham and Swansea was instigated on this service, with the locomotive subsequently working on to Carmarthen with a following train. This train ran via the North Warwicks line with 'A' lamps to Cardiff, and 'B' onwards to Carmarthen, reaching the latter town at 3.30 p.m.

Another view looking west from the viaduct bank, this time showing Old Oak-based No. 5038 *Morlais Castle* at the head of the 8.48 a.m. Fishguard to Paddington parcels train, probably c. 1941. This train conveyed mostly parcels vehicles from Fishguard, Carmarthen, Llanelly, Swansea and Neath to Cardiff, Swindon or Paddington, though vans for Sheffield, Nottingham and the LMS line were also attached. Vehicles from Aberdare and several other locations in the Valleys were added at Cardiff along with traffic from Cardiff itself, with around 17 vans, brake vans, 'Siphon Cs' and 'Siphon Gs' scheduled into Paddington. On Saturdays, the train conveyed passengers between Swansea and Cardiff, hence the 'B' headlamp. The train was scheduled to return most of the vehicles from the early morning newspaper train (12.55 a.m. Paddington), and would have been conveying the Fishguard (TThSO) and Swansea vehicles at this point, picking others up at Cardiff and Newport. During the mid- and latter 1930s, engines from Swindon shed were regularly seen on this train. During the earlier part of the war, Old Oak 'Castles' were rostered to work the 8.30 p.m. Paddington to Neyland parcels to Swansea, returning to Paddington with the 8.48 (via Gloucester). In postwar years, the train always changed engines at Cardiff, and was worked forward by a Gloucester engine as far as Swindon, mostly in the early 1950s by No. 4059 *Princess Patricia*, and, following her demise, by 5017 *The Gloucestershire Regiment 28th–61st*; the working into Cardiff at that time was by a Neyland 'County' 4-6-0.

In this photograph, Canton 'Castle' No. 5080 *Defiant* is seen heading the 8.15 a.m. Neyland to Paddington service, probably c. 1941. No. 5080, built in May 1939, and seen here with the plated-in cab window, was renamed from *Ogmore Castle* in January 1941. The engine was probably working a Cardiff turn involving four trips to or from Swansea, starting with the 7.35 a.m. Cardiff (for Neyland), then the 8.15 a.m. Neyland, the 1.55 p.m. Paddington, and finally the 9.5 p.m. Swansea, amounting to about 190 miles in all. In summer 1941, the train was scheduled for five vehicles (including the diner) from Swansea, then five from Neyland and three from Pembroke Dock at the rear in this instance formed with a rather pristine set of coaches,

Following the cessation of regular double heading on heavy trains to the west of Cardiff after the introduction of 'Castles', the use of a pilot was not frequent, other than to avoid light running. This late 1930s view shows one occasion when the 11.20 a.m. Milford Haven to Paddington (Train No. 835) was so formed, with Oxley 'Hall' No. 5966 *Ashford Hall* piloting the normal 'Castle' as they eased off for the 60 m.p.h. restriction around the curve under the viaduct. On up trains, the number was carried on the smokebox door from Swansea to Paddington, but the down Welsh services were not numbered at this time.

This pastoral scene shows the penultimate engine of the '28XX' class, No. 3865 (built November 1942), drifting around the curve towards the Barry viaduct with an up class 'H' freight in 1946. No. 3865 was one of the first in the batch of ten '28XX' class 2–8–0s to be converted for oil burning at a time of acute coal shortage, entering traffic as such in December 1945. Allocated to Severn Tunnel Jct after conversion, the engine was renumbered 4851 in December 1946. A further ten were converted in the summer of 1947, though from the latter part of 1948 a shortage of foreign currency to purchase the oil caused the reversion to coal burning. From the 1930s, South Wales had become increasingly an industrial area, and the movement of coal (for which there was less demand, and a postwar shortage in any event) played an ever-diminishing (though still very important) part in the South Wales railway scene. In 1929, the Great Western had conveyed some 45 million tons of originating coal and coke, a figure that had reduced to 22 million in 1946 (the vast majority still from South Wales).

A 'Hall' class 4–6–0 heading an up express to the east of St Georges one late morning, c. 1941. The engine may well have been Reading's No. 5948 *Siddington Hall*, which was known to work in South Wales around this time, being 'stopped' for repairs at Cardiff in February 1942, and having a tender change at Newport the following month. On this occasion, it was in charge of a 13-coach train (possibly a division of the 8.15 Neyland to Paddington). No. 5948 was much used on the Reading & Portsmouth and Reading, Henley & Paddington trains around this time, but also travelled further afield with freight or troop train workings, an everyday part of the class's duties then. The former Barry Railway line from St Fagans to Pontypridd can be clearly seen in the top right-hand part of this picture as it climbed away towards Tynycaeau Jct.

Unlike many 'foreign' through services, the 10.50 a.m. (Sundays) Swansea to Sheffield train continued to run during the war years, though its weekday counterpart (8.15 a.m. Swansea to Newcastle) was suspended for the duration and not reintroduced until 1946. During the war, the LNER stock was frequently observed in use on weekdays for local duties in South Wales, and the four-coach corridor set (specified Brake Third, Compo, Third, Brake Third, plus strengthening stock) is seen here behind Swindon-based No. 2913 *Saint Andrew* on a 'B' lamp local service, probably in the early war years. Although passenger-rated brake and other vans from the other three companies were a common sight on much of the peacetime Great Western (as indeed were passenger-carrying vehicles), their use was fairly strictly regulated. In wartime, and in the immediate postwar era, these conditions seem to have been relaxed, and surviving records show the use of 'foreign' passenger coaches on Paddington and other trains in addition to their more regular duties. No. 2913 was condemned in May 1948.

A service of auto-trains – five daily (six on Saturdays) each way in 1939 – ran between Pontypridd or Tonteg and Cardiff via Creigiau and St Fagans, with a '64XX' class engine and a couple of auto-coaches normally forming the trains. These normally terminated at Clarence Road, though the last return service of the day ran to and from Riverside. Originally, they ran from the Barry station at Pontypridd Graig (closed in 1930), and subsequently from the ex-Taff Vale. In this view, the auto service is seen on the bank between Tynycaeau Jct. and St Fagans (singled in 1934), from where the services utilised the South Wales main line, in the late 1930s or early war years.

The ex-Barry Railway main line between Cadoxton Jct. and Trehafod Jct. was carried over the South Wales main line by the imposing six-arch, 180-yard St Fagans Viaduct. In the postwar era, regular passenger services running between Pontypridd, Porth, Treherbert, Aberdare and Barry, and numerous mineral trains between the Rhondda, Aberdare and Rhymney Valleys collieries and yards and Cadoxton, used the viaduct, mostly hauled by 0–6–2Ts. In contrast, 'Castle' 4–6–0 No. 5029 *Nunney Castle* from Old Oak had just passed under the viaduct at the head of the 9.5 a.m. Bristol to Swansea c. 1946, the engine probably running onwards from Cardiff.

Another oil burner, No. 2872 (from Llanelly) passing under the viaduct at St Fagans with a down class 'H' freight early one afternoon c. 1946. The mixed load evidently included two tank wagons, three wagons of what could have been pit props near the front, and a large number of sheeted opens. At this time, around ten of the oil burners were allocated to South Wales sheds, and could regularly be seen on the main line through St Fagans. The flow of through heavy freight trains in the down direction at this time was largely from Severn Tunnel Jct or Rogerstone to the Neath, Swansea or Llanelly areas, though trains from Banbury and Gloucester also ran through daily. On the other hand, eastbound traffic passed in some quantities direct to Saltney, Bordesley, Stourbridge, Gloucester, Banbury, Moreton Cutting and destinations near London, as well as to Newport and Severn Tunnel Jct. The signals above the engine on the ex-Barry were for Tynycaeau South box, just to the north of the viaduct, which

Except on a few vacuum services, and on balancing turns in such places as Cornwall, the regular use of 'Castles' on freight work was not very common. This picture illustrates Landore-based No. 5013 *Abergavenny Castle* with a lengthy down 'K' lamp train on the curve under the viaduct, almost certainly during wartime, and probably on a balancing turn to put the engine to some use on the return to its home shed. As Group 'D' engines, their permitted heavy freight loads equated to those of '49XXs' and '68XXs', although they were rated alongside the '47XXs' for vacuum trains. In later years, the 8.55 a.m. Cardiff (Newtown) to Llandilo Junction was often diagrammed for a 'Castle' as the last leg of a return from Paddington.

Entering traffic in September 1944 at Severn Tunnel Jct., 'Hall' No. 6969 was not named *Wraysbury Hall* until June 1947, at which time it moved on to Canton shed. Hubback photographed the engine in 1946 at the head of the 11.15 a.m. Cardiff to Neyland parcels train under the viaduct at St Fagans. This service conveyed regular vehicles from Birmingham, Wolverhampton, Sheffield, Bristol and Cardiff to Fishguard, Whitland or Neyland, avoiding Swansea by running via Felin Fran and the District line. The use of the Forest of Dean wagon in the heavy freight on the up line illustrates the pool arrangements still in force, whereby private owner vehicles from all parts of the country were widely used for traffic all over the Great Western system, including minerals in South Wales.

No. 5096 *Bridgwater Castle* passing under St Fagans viaduct with a down express, possibly the 8.55 a.m. Paddington to Pembroke Dock, watched by a group of young boys sitting on the bank and at least partly occupied in catching butterflies. Records show that No. 5096, from Bath Road shed, was more at home on the road between Bristol and Paddington, though she was also regularly used on the through

A fine study of Canton 'Saint' No. 2905 *Lady Macbeth* easing round the curve under the viaduct with the 9.5 a.m. Bristol to Swansea. In this instance, the train conveyed van traffic at the head, probably destined for Swansea. Tail traffic was often so positioned to be dropped off en-route, where the receiving station had a locomotive to deal with it; however, if the station did not have motive power to hand, the vehicle would need to be positioned by the train engine, and would therefore probably be at the front of the train. No. 2905, possibly seen here in wartime black, was condemned in April 1948 after 42 years' service.

No. 4094 *Dynevor Castle* heading west under St Fagans viaduct with a Paddington to West Wales express, probably the 8.55 a.m. to Pembroke Dock, in the postwar era. The engine was transferred to Canton from Laira in November 1937 (via Swindon Works), and remained at Cardiff until late 1952. No. 4094 was recorded throughout the 1940s on numerous occasions at Paddington with the 8.15 a.m. Cardiff and 1.55 p.m. Paddington return, and on the 8.15 a.m. Neyland with the 5.55 p.m. return. With their through sections to and from Swansea, Carmarthen, Whitland, Pembroke Dock, Neyland, Milford Haven or Fishguard, the postwar Welsh expresses were closely assembled, their present counterparts are then did the West Country or Northern expresses.

The 9.5 a.m. Bristol to Swansea was frequently hauled by 'Saints' in the 1940s. This train worked as a 'fast' from Bristol, calling at Stapleton Road, Pilning (to pick up), Severn Tunnel Jct., Newport and Cardiff, and thereafter as a stopping train to Swansea. No. 2980 *Coeur de Lion* (from Canton) is seen on the turn in this instance, working on from Cardiff, probably c. 1941. Two years later, the turn was handled by Landore shed, which worked the 7.40 a.m. (to Gloucester) as far as Cardiff, returning to Swansea with the 9.5 Bristol, then on to Carmarthen with the 1.55 p.m. Paddington, and finally back to Swansea with the 7.25 p.m. Neyland Perishables.

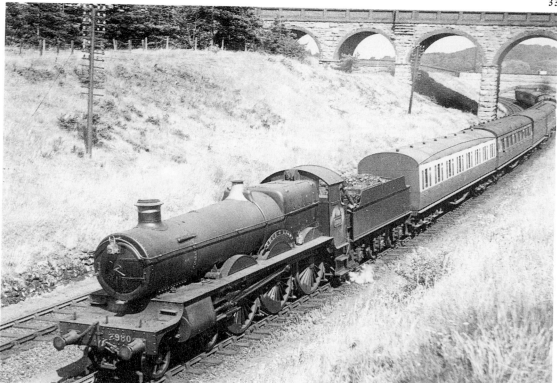

Between September and December 1940, a batch of 25 LMS class '8F' 2–8–0s were transferred onto the Great Western, and allocated to Severn Tunnel, Ebbw Jct., Cardiff, Neath, Landore and Llanelly sheds. In January 1941, seven of these were returned, as were the remainder between July and October of that year. No. 8240 worked on the GWR from October 1940 to September 1941, based at Llanelly, and was photographed at the head of a down 'F' class mixed freight at St Fagans viaduct in 1941. The headlamps, a very different design to their GW counterparts, are seen with extended shades.

Hubback photographed St Fagans viaduct from all angles, but unfortunately never took anything passing over the top. In this view, the structure is seen from the east side, from where the lower accommodation bridge can also be clearly seen. The latter bridge was in two sections, spanning the main lines (left), and, at a higher level, the loop to Tynycaeau Jct. (right). Canton 'Grange' No. 6827 *Llanfrechfa Grange* can be seen passing with an up coal train, c. 1941, carrying Canton target 'H2', which covered the 2.50 a.m. trip from Cardiff (Penarth Curve) to Tondu and the late morning return journey. The engine was still quite clean, with the plated cab window and 'shirt button' emblem on the tender.

A leaking Canton 'Castle' No. 5001 *Llandovery Castle* heading an up express, probably the 8.15 a.m. Neyland to Paddington, under the bridges at St Fagans. Taken in the early years of the war, the photograph shows the cleanliness of the engine despite the prevailing conditions. Although station nameboards had been reduced in size, or even removed to 'confuse enemy parachutists', the main trains continued to carry roof label boards during most (if not all) of the war. Restaurant cars were still provided on the main Welsh trains in 1941, though the number had dwindled to just two each way by 1943, being withdrawn completely in 1944.

Old Oak Common-based No. 5027 *Fairleigh Castle* passing under St Fagans viaduct with train No. 830, the 10.20 a.m. Pembroke Dock to Paddington, in the late 1930s. No. 5027 entered traffic at Old Oak in May 1934, and remained there until September 1950, when it moved to Chester (via Swindon Factory). As well as the usual West Country services, the engine was used regularly on South Wales trains, and was recorded on arrival at Paddington with the 10.20 a.m. Pembroke Dock on Saturday, 12th August 1939, at the head of 14 coaches.

A view from the accommodation bridge at St Fagans, showing Canton-based No. 5065 *Newport Castle* (briefly *Upton Castle*, when built) approaching the viaduct with the 11.55 a.m. Paddington to Milford Haven in the late 1930s. The formation of the train was scheduled as four coaches for Swansea (including the dining car), four for Milford Haven, one for Fishguard Harbour, and three for Pembroke Dock at the rear. On peak Saturdays during the summer, the train was divided, being preceded by a service to Neyland (11.50 Paddington). On Saturdays in August 1939, the 11.55 was also strengthened by the addition of extra vehicle(s) at the head of the Pembroke section, as may have been the case here. The former Barry Railway line descending from Tynycaeau Jct can be seen on the left.

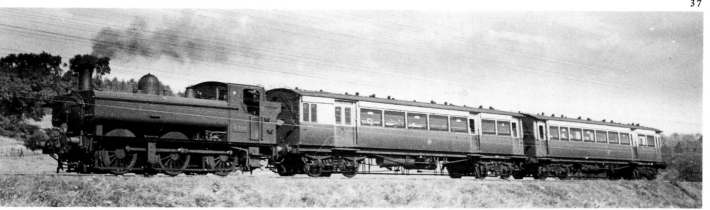

The Cardiff (Clarence Road) to Pontypridd auto on the single line from St Fagans to Tynycaeau Jct., photographed in the early 1940s from the south side of the main line, and worked by Abercynon 0–6–0PT No. 6411 with just a hint of 'Great Western' painted on its tank side. In wartime, the service was reduced to four trains each way, from the five (six on Saturdays) of the prewar timetable. These workings continued until the line closed in the 1960s. Coaches were ex-Taff Vale, with an Intermediate Trailer Third and Driving Trailer Compo forming the train. From St Fagans, the line ran parallel with the main, climbing through the viaduct arch before turning north to Tynycaeau Jct to join the Cadoxton & Treforest Jct line for Pontypridd.

Canton 'Saint' No. 2906 *Lady of Lynn* survived until August 1952, a working life of over 46 years. Shortly before this, she was specially cleaned up, and is seen working a Cardiff to Tondu train on 26th July 1952, photographed from the accommodation bridge near St Fagans viaduct. This engine was transferred to Canton in 1935, and became a familiar sight on the South Wales main line, though for the most part on secondary expresses and other passenger turns.

St Fagans station was set in a most attractive area, with bridges over the Ely River and a level crossing on a long, gentle curve of track. In this view, an early 'Castle' (possibly No. 4085) was working a down London express through the station one evening in 1931/2. Dean's smaller passenger brake vans were still a common sight at this time, and one example may be seen behind the tender; the 3.55 p.m. Paddington to Carmarthen was scheduled to convey a van destined for Fishguard, and this may be the train portrayed. The South Wales Railway company's broad gauge line was opened between Chepstow and Landore in June 1850, and a through connection from Chepstow to the GWR line at Grange Court two years later. St Fagans was served by just three passenger trains each way in the early days, but these had increased sixfold by 1910 (half of which were provided by the Barry's Pontypridd & Cardiff service).

Canton-based No. 5012 *Berry Pomeroy Castle* passing through St Fagans station with the 1.55 p.m. Paddington to Neyland train in the late 1930s. The front vehicles were scheduled for Swansea, whilst the four-coach Neyland section is that roof-labelled at the rear, ahead of the tail traffic. The engine roster for the mid-1930s shows a Cardiff engine working to Paddington with the 8.15 a.m. Cardiff, returning with the 1.55 p.m., possibly as far as Swansea, which may have been the working shown here. No. 5012 was at Cardiff between 1933 and 1939. The goods yard and cattle pen, seen on the left, were served by a single siding trailing from the up main at the east end of the station.

Although the 'Hall' class were more popularly associated with passenger duties, in practice an equal amount of their work was done on freight turns, and more so in industrial areas. Canton-based No. 4913 *Baglan Hall* is pictured here with the Canton 'H17' target, the 3.50 p.m. Cardiff (Penarth Curve) to Tondu empties c. 1946. Group 'D' engines, such as the 'Halls', could haul 100 empties on the easy grades of the section, but were restricted to 76 (unassisted) up the 1 in 106 gradient between Llantrisant and Llanharan. The engine was still carrying the wartime plated-over cab windows, the original glass, together with its beading, being replaced by a similarly-sized sheet of metal.

'Star' No. 4064 *Reading Abbey*, newly transferred to Landore, heading the 8.0 a.m. Pembroke Dock to Paddington express on the approach to Ely on Wednesday, 4th July 1934. The first three coaches of the train (including the dining car) were from Swansea, the next three from Pembroke Dock, and the following two from Neyland. The vehicle at the end was scheduled as a 'Siphon G' from Landovery, though it appears to have been a passenger brake or similar vehicle. The engine did not remain in South Wales for long, being transferred away in October 1935. In February 1937, No. 4064 was withdrawn from

ELY WEST

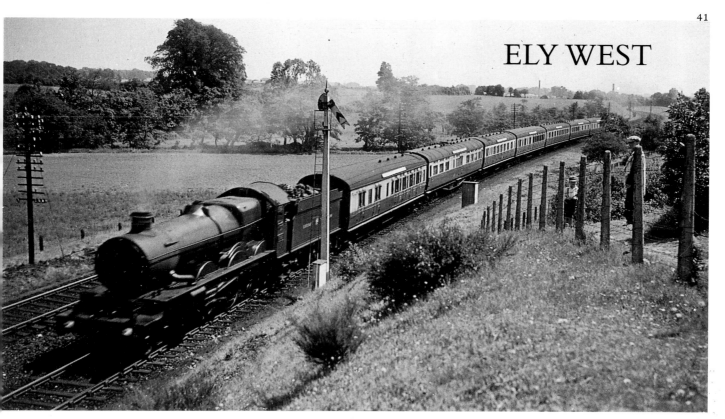

The 8.55 a.m. Paddington to Pembroke Dock express passing St Fagans down distant signal behind No. 5029 *Nunney Castle*, of Old Oak. The date is recorded as 4th July 1934, when the engine was a matter of weeks old, having recently emerged from Swindon Factory and Tyseley Shops. As with 13 other 'Castles' introduced between 1934 and 1938, No. 5029 remained an Old Oak engine throughout her Great Western days, whilst a few more of that era did likewise at other sheds. Her train carried the usual vehicles for Swansea (3), Pembroke Dock (3) and Neyland (2).

'Saints' were still active on the main London expresses within South Wales until the mid-1930s. Canton acquired No. 2990 *Waverley* in late 1932, and retained it until withdrawal in January 1939. The engine was pictured here passing St Fagans down distant on its way to Cardiff with what is believed to have been the 10.20 a.m. Pembroke Dock on Saturday, 17th August 1935. In its introductory year, 1934, the train numbering system covered only West Country services, and, in the case of South Wales, trains to or from Weymouth, Paignton and Plymouth carried identification. In 1935, up South Wales trains to Paddington were included in the scheme (in the 8XX series).

Another express between Ely and St Fagans with No. 4089 *Donnington Castle* heading west with a down service c. 1936. No. 4089 was transferred to Landore in late 1935 for a period of about two years, during which time most of her duties would have involved the London trains. Clerestory vehicles were still quite common in secondary main line trains, and two may be seen in the centre of this train in the process of crossing the Ely River, which accompanied the line from the outskirts of Canton westwards to Llantrisant station.

'Star' No. 4059 *Princess Patricia* was a Landore engine between 1932 and 1938, having previously spent some time at Carmarthen shed. She is seen here heading the 4-coach through portion of the 7.5 a.m. Kingswear to Swansea service passing over the River Ely bridge about a mile to the west of Ely station at 1.25 p.m. on Wednesday, 4th July 1934. The earliest of the Saturday morning trains from the West Country did not carry reporting numbers; so far as South Wales was concerned, the only one was the 11.10 a.m. Paignton, and then only as far as Cardiff. This formation was scheduled to return the following morning attached to the 9.5 a.m. High Street to Cheltenham train as far as Cardiff, forming the 11.25 a.m. thence to Paignton.

CANTON AND CARDIFF GENERAL

One mile to the west of Cardiff General station, the original main line at Leckwith Junction made a sharp dog-leg to the left beyond Canton Loco prior to the remodelling of 1934. Hubback recorded this view, looking west from Canton footbridge, one afternoon in December 1931, clearly showing the old deviation. Canton Loco is seen on the left, and an interesting selection of coal, cattle and general merchandise vehicles in Canton Mileage Yard, on the right. The end-loading ramp is seen at the throat of the yard.

At the same time, he recorded an up express, hauled by Canton's 'Saint' No. 2905 *Lady Macbeth*. The effect of the curve can be seen as the rear coaches of the train negotiate the final part of the deviation. Although coach roofs were painted white at this time, the effects of weathering is evident here in comparison with the gleaming roof of the penultimate vehicle in the train.

This view from the new Canton footbridge was taken post-1934, after the deviation at Leckwith had been eased, though the track still curved gently to the right in the distance to connect with the original alignment beyond. No. 2971 *Albion* is seen heading an up stopping train on the new layout. The connection from the yard to the up main had been removed in the remodelling work, and the end-loading dock had been moved to the siding on which the horsebox was standing. Beyond the train, the Leckwith Road and Grosvenor Street level crossing had been replaced by new bridges.

A view eastwards from Canton bridge, with Canton 'Castle' No. 5012 *Berry Pomeroy Castle* hauling a down express in the latter '30s. The long stone building beyond the locomotive was the old carriage shed, destined to become a milk and fruit depot in the 1934 modernisation, with access provided from de Croche Place. A new 11-road carriage shed to accommodate 150 vehicles was constructed to the south of Canton Loco, and a new footbridge, 436ft long (from which this photograph was taken), connected it with de Croche Place.

Pictures of Cardiff General before the remodelling of 1934 are not common. This early 1930s view at Cardiff East shows the 8.55 a.m. Paddington to Pembroke Dock approaching the down platform behind a 'Castle', with sections for Swansea (including the dining car), Pembroke and Fishguard. This train had slipped a Weston-super-Mare portion at Stoke Gifford some fifty minutes earlier. The Southern stock behind the Pembroke train was for the 12.20 p.m. through service to Brighton via Salisbury and Chichester, which was due to arrive there at 6.30 p.m. Although a quicker journey to Brighton was possible via Paddington and Victoria, the train did provide the convenience of a through service, good connections to numerous points in the south of England, and a dining car during the summer months. The frontage of the Central Hotel is seen on the left, with its prominent advertising. The positioning of the semaphore signalling was quite different to the colour light system that replaced it, as can be partly seen by reference to the picture below.

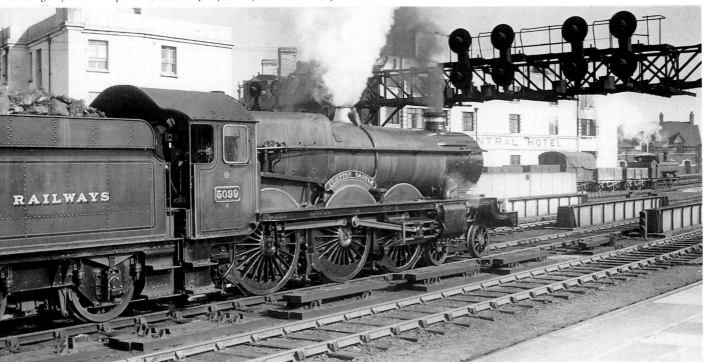

Colour light signalling was provided throughout the Cardiff General station area under the 1934 remodelling, and was concentrated on gantries on the bank between Newtown West and Cardiff East for the down direction, and straddling the tracks from the ends of Platforms 2 and 3 for up trains at Cardiff East, with a small gantry at the end of Platform 1. This picture, taken in the late 1940s, shows part of the layout provided in 1934. Canton's prize engine, No. 5099 *Compton Castle* (claimed by the Canton shedmaster to be the best 'Castle' ever built), is seen standing in the middle road awaiting an up London express, which it would work on to Paddington. The open cab pannier in the background was working an up transfer freight between the Cardiff yards past the modernised frontage of the Central Hotel.

One of Hubback's venues in postwar years was at Gaer Junction, photographing trains as they emerged from or approached the south ends of the twin High Street Tunnels at Newport. There was a considerable range of semaphore signals at this busy junction, and those situated at the two portals controlled all four running lines. Reading-based No. 5956 *Horsley Hall* is seen emerging from the New (Western Valley) Tunnel with an 'E' class train of empties bound for the Rogerstone line; this engine was more commonly used on passenger trains between Reading and Paddington in the postwar period. The new tunnel was opened in 1911 in conjunction with the quadrupling of the track from High St. station to Ebbw Jct., to connect with the existing quadruple track from there into Cardiff.

NEWPORT, GAER JUNCTION

A run-down Westbury 'Bulldog', No. 3363 *Alfred Baldwin*, making hard work of restarting an up freight from the Cardiff direction in 1946. Judging by its condition, the engine was probably confined largely to freight work, but was not withdrawn until October 1949, some three years after this view was taken. The engine was about to enter the Old Tunnel on the up main, probably taking the up through road at Newport station, at the east end of the tunnel. At this time, there were still around fifty 'Bulldogs' in service, though the last was withdrawn some five years later, in 1951.

Oil burning '28XX' No. 3818 at the portal of the 742-yard Old Tunnel with the signals 'off' for a clear run through Newport station with an up main line freight in 1946. This engine was converted for oil burning in January 1946 and transferred to Severn Tunnel Jct., being renumbered 4852 in December of that year. Oil fuel was fed by gravity to a burner located in the front of the firebox, and atomised by means of a steam jet incorporated into it. The tender carried 1,800 gallons of fuel, with operational refuelling depots at Severn Tunnel Jct., Newport, Cardiff and Llanelly at that time. No. 3818 was reconverted for coal burning in September 1948.

Eighty Stanier '8F' 2–8–o designs appeared from Swindon works between 1943 and 1945, and were utilised for heavy freight duties alongside the '28XXs' on the Great Western system. The class were to be found at sheds between Old Oak, Laira, Cardiff, and Birkenhead, with concentrations at Oxley and St Philip's Marsh. This picture shows Oxley's No. 8478 emerging from the 'Western Valley' Tunnel on the Down Relief in 1946, bound for Park Jct. and the Rogerstone line. This engine was in GWR stock from June 1945 until September 1947 when it was sent to the LMS.

Another Stanier '8F' on the GWR, Pontypool Road's No. 8468 starting away with an up freight towards Old Tunnel while Ebbw Junction's '45XX' No. 5516 was waiting inside the mouth for the road. Around this time, the '45XXs' were recorded on the Ebbw Jct. Pilot or Banker and station pilot turns, as well as on passenger duties on the Blaenavon and Brecon routes.

A Stanier '8F' (with 'STJ' stencilled on the frame above the cylinder) on the Up Main alongside Gaer Jct. box restarting a mixed freight, which included coal traffic conveyed in the usual wooden opens. With the level of traffic through Newport High St. station, many freights were brought to a stand outside the Junction box awaiting the road. Some of these had previously ground to a halt (often for a considerable period) at Ebbw Junction, half-a-mile to the south, where crews might be changed and engines watered.

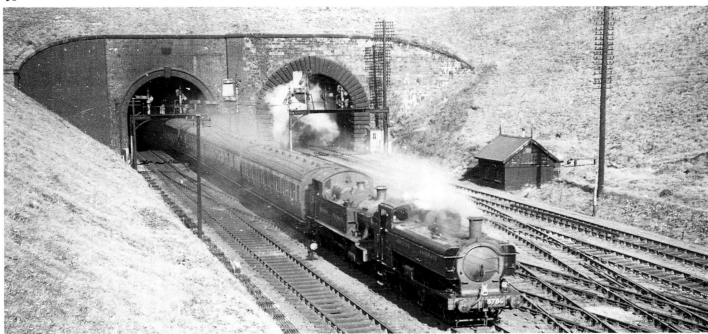

An interesting combination of motive power was provided for this passenger train, seen emerging from the New ('Western Valley') Tunnel behind '57XX' class pannier tank No. 8786 and a '45XX'. Both classes were common on passenger turns around Newport, the '57XXs' particularly on the Blaenavon, Aberbeeg and Brynmawr services. Although the '45XXs' were more abundant in the West Country, within South Wales divisions around this time they were to be found at Pontypool Road, Newport Ebbw, Aberbeeg, Aberdare, Canton, Tondu, Whitland and Pembroke Dock sheds.

The 9.45 a.m. Birmingham to Cardiff express emerging from Newport (Old) Tunnel behind Hereford's well-groomed 'Saint' No. 2932 *Ashton Court*, with all signals off for a clear run on the Down Main. Beyond Gaer Jct. box, the 'Saint' would swing to the right across the 'direct running junctions' of the main line (speed limit 30 m.p.h.) to Cardiff, whilst the road on which it was photographed accompanied the Main, to its east side, as the Goods (Relief, beyond Ebbw Jct.). Clerestory stock was occasionally in use on such services at the time, though it was becoming increasingly rare.

Until the 1960s, two of the more interesting workings into Newport included the Sirhowy Valley services and through trains to and from Brecon. In the 1940s and early 50s, the Brecon trains were mostly worked by Ebbw Junction and Brecon '2251' and '2301' Class 0–6–0s, but sometimes produced a Cambrian 0–6–0 from Oswestry, all being finally superseded by LMS/BR 2–6–0s of the '465XX' series. LMS Webb 'Coal Tanks' from Tredegar worked the Sirhowy Valley service with stock that was often in the vintage class, and sometimes cascaded from other areas, hauled here by Tredegar's No. 7834 in 1946. These were superseded in the mid-50s by Ivatt 2–6–2Ts, until the service was cut short at Risca and worked by WR auto services. The train is seen joining the Up Relief from the Gaer Loop line from Park Jct., with the main lines over to the left of the engine. Beyond the train, the identification of running lines changed, with the Goods lines to the left and the Main lines to the right, with the Main 'direct running' crossover for passenger trains between them; the 'direct running' ('high speed') connection allowed a 30 m.p.h. passage as opposed to 15 m.p.h. over the adjacent scissors crossover.

Newport 'Bulldog' 4–4–0 No. 3422 is pictured on the 5.15 p.m. Newport to Gloucester on 15th June 1931 at Lliswerry, just east of East Usk Yard, Newport. There were about two dozen of the class allocated to the main line sheds between Severn Tunnel Jct. and Neyland, with a particular concentration in the western section. In South Wales, they were mainly found on secondary express or local passenger duties, which also brought sister engines from Gloucester and Bath Road into the Principality. Although there had been an urgent need for quadrupling between Newport and Severn Tunnel Jct. since the Great War, the work was not completed until 1941. A 1920 official report had suggested that, given traffic density, quadruple track was needed throughout between Severn Tunnel Jct. and Pembrey.

The Severn Tunnel Car Train, which ran between Severn Tunnel Jct. and Pilning stations, is seen here descending towards the western portal of the tunnel behind 2–6–2T No. 4121. In 1946/7, the train ran twice daily (three times on Mondays and Saturdays), with small cars charged at 7s and larger vehicles at 8s 5d, waterproof sheeting extra. This service was operating in the early 1920s, though carriage vehicles were then attached to ordinary trains calling at Patchway or Pilning stations. The grimy state of the engine was typical of the condition of the Severn Tunnel 2–6–2Ts, used for piloting freight (and occasionally passenger trains) through the tunnel. Because of the nature of the work involved, there was little point in cleaning these engines, which seemed to spend much of their time in the grime of the tunnel. Their external condition was not, however, a reflection of their general mechanical state of repair, which was very good. No. 4121 was one of the batch of ten 1938-built '51XXs' that spent virtually all of their days at Birkenhead during the Great Western era, the engine moving to STJ shed soon after nationalisation.

The line from Severn Tunnel Jct. entered the Severn Tunnel about a mile to the east of the station, swung south-eastwards under the Gloucester line around ¼-mile later, and passed under the western shoreline of the river a mile further on from that. This view shows Landore-based No. 7003 *Elmley Castle* completing its climb out of the tunnel, having just passed Severn Tunnel West Box. Built in June 1946, this engine spent much of its life working from Landore, and was a mainstay of the postwar turns to Paddington, though the number of LMS vehicles in her train on this occasion suggests a North & West service.

This picture of Stafford Road's 'Star' No. 4053 *Princess Alexandra*, c. 1950, illustrates that the class were still performing important work in their final years. Heading through Pilning down the gradient towards the Severn Tunnel, she was in charge of a relief train, probably for the North. The coaches on the up refuge siding were part of the Severn Tunnel car train. In 1942, due to pressure of traffic, the down goods loop to the east of Pilning Jct. was extended past the station and on to Severn Tunnel East, Ableton Lane Tunnel, giving a 1½-mile length. The new up loop, constructed at the same time, ran for 1 mile from Ableton Tunnel to Pilning station. More than one goods train could be allowed into the loops at the same time under the permissive block system.

A '43XX' 2-6-0 on the Saturdays-only 11.45 a.m. Portsmouth & Southsea to Cardiff (due 4.25 p.m.) at Patchway, c. 1950. Much use was made of these engines on summer Saturday schedules with cross-country expresses, normally with around eight to ten coaches, as seen here. On the Southern Region through trains via Salisbury, the engines were mostly supplied by Canton and Bath Road, though at peak periods locomotives from other sheds were also involved. Running through unassisted from Salisbury, the '43s' were allowed a maximum of 360 tons on the climb to Warminster, a load at the limit of their capability to keep time, albeit on the rather slower schedules. Where late running was experienced, as was very often the case on these trains to South Wales, use of ageing '43XX' class engines sometimes caused a chain reaction to following services.

PILNING AND PATCHWAY

In December 1948, three 'Kings' (Nos. 6000, 6018 and 6019) were transferred to Bath Road shed, and were regularly recorded on Bristol or Weston expresses to and from Paddington. Nos. 6018 and 6019 only remained there until July 1950, but No. 6000 *King George V* stayed until October 1952, and became involved on through workings between Bristol and Shrewsbury with North & West expresses. The engine is seen here with Train No. 593, the 12.15 p.m. Kingswear to Manchester (4.30 p.m. Temple Meads), passing through Patchway c. 1950.

Cheltenham Spa, Lansdown Jct., with Swindon '43XX' No. 6374 departing from the ex-Midland station with the 1.55 p.m. Cheltenham to Southampton train (via the M & SWJ line) in postwar years. Prior to grouping, the M & SWJ trains had used Lansdown station, an arrangement that the GWR and BR (Western Region) continued; doubtless, this was more convenient for passengers travelling between the North, the Midlands and Southampton, requiring only a change of trains at Lansdown. The Western Region route to Malvern Road station, Stratford-on-Avon and Birmingham ran straight ahead at this junction, the line to St James' station turning off to the right some ¾-mile distant, beyond Malvern Road. '43s' had joined the familiar 4-4-0s on the ex-M & SW line by the mid-1930s, mainly on goods runs, but during the latter years of the war they had taken over nearly all services.

CHELTENHAM

The 9.25 a.m. Swansea to Birmingham Snow Hill (10.58 a.m. Cardiff) at Cheltenham Lansdown Jct. in the early 1950s, with a resplendent Landore 'Castle', No. 4095 *Harlech Castle*, in charge. The ex-GW Banbury & Cheltenham line diverges to the left in front of the box, along which traffic for the ex-M & SWJ for Swindon and Southampton also passed as far as Andoversford Jct. The line between Gloucester and Cheltenham was quadrupled during the Second World War to cater for the ever-increasing traffic of the two companies using the section.

The LMS line through Cheltenham carried traffic between the North of England (ex-Midland section) and Bristol, where some through trains and individual vehicles were exchanged with the Great Western at Temple Meads. A double-headed service between Cheltenham and Gloucester is seen here piloted by a well-presented Johnson/Deeley 4–4–0 '3P' No. 716 (Saltley) and an LMS 4–6–0 'Class 5', carrying reporting No. W212, in the mid-1930s.

Gerald Hubback travelled at weekends between Taunton and his home at Barry, often using P. & A. Campbell's steamers across the Bristol Channel between the harbours at either Weston or Minehead, and Barry. On these occasions he took photographic studies at both places, which afford an interesting record of workings during the summer of 1932. Here, a curious double-headed arrangement with a curved-frame 'Bulldog' (possibly No. 3325) and a '43XX' is seen on the eastern approach to Weston-super-Mare station, leaving the main line for the Locking Road platforms with empty stock for a return excursion.

WESTON-SUPER-MARE AND MINEHEAD

Prime power in the form of Old Oak 'King' No. 6029 *King Stephen* was provided for this heavy 'Kiddies Express', approaching Weston-super-Mare station, with many eager faces at the windows. The 'Kiddies' trains followed the introduction of another 'novelty' service, the 'Hiker's Mystery Express', in 1932. Despite their use on the heaviest expresses on the Northern and West Country routes, the 'Kings' often appeared on such excursions as these, and also on the special trips to Swindon Works. An interesting set of vintage stock can be seen in the siding, probably having worked in with another excursion. No. 6029 was renamed *King Edward VIII* in 1936.

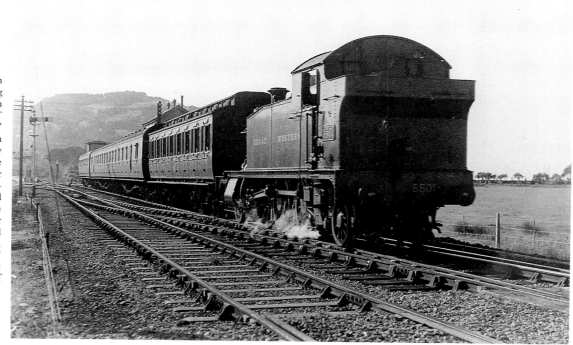

Minehead station on 30th June 1934, with an evening train departing for Taunton behind '45XX' class 2–6–2T No. 5501. A 2-coach 'B-Set' had been strengthened at both ends, with an ageing clerestory Third at the front. The Whittaker apparatus (for exchanging tokens automatically at a greater speed than allowed by hand) attached low on the bunker side of the engine, was found on many of Taunton's 4–4–0, 2–6–0, 0–6–0 and 2–6–2T allocation. The 1½-mile section to a point just west of Dunster was double track.

DURSTON

'Star' No. 4035 *Queen Charlotte* (from Bath Road shed) at the head of the 7.30 a.m. Paddington to Kingswear (Saturdays – Train No. 100) in the cutting to the south of Durston station during the late 1930s. This was a rather slow service, running via Bristol, and calling at Reading, Didcot, Swindon, Chippenham, Bath, Temple Meads, Bridgwater, Taunton, Exeter, Teignmouth and all stations to Kingswear, with a scheduled journey time of six hours and ten minutes, over two hours more than the 'Torbay'. The front half of the train was formed with Bristol to Paignton coaches, and the rear by the Paddington to

COLE

Hubback's stamping ground during the time he worked and lodged at Taunton during the 1930s stretched from Cole to Exeter. Cole was one favourite location, where this picture of Newton 'King' No. 6024 *King Edward I* was taken in the summer of 1937 as it approached the S & D line bridge. Train No. 605 was the Saturdays-only 9.0 a.m. Perranporth to Paddington, which, in addition to conveying the Perranporth and Truro sections, called at Brent to attach the Kingsbridge through coaches, seen here at the front (Van Third and Compo scheduled). This train was very restricted in its station calls en route, stopping only at St Agnes, Truro, Liskeard and Brent, thence non-stop to Paddington.

The Somerset & Dorset main line crossed the GWR's former 'Weymouth Line' near Cole, between Bruton and Castle Cary, by means of the girder bridge seen in this view. No. 5017 *St Donats Castle*, a Newton Division engine in the latter half of the 1930s, is shown passing under the S&D line with Train No. 142, the 10.35 a.m. Paddington to Penzance service, c. 1937. This summer Saturday train carried vehicles for Falmouth or Helston in addition to the main Penzance section. In 1935, the 10.35 a.m. Paddington train was titled 'The Cornishman' serving Weymouth (slip), Newquay, Falmouth and Penzance each weekday, and St Ives, Falmouth and Penzance on Saturdays. No. 5017 left Newton in February 1939 for Swindon Factory, after which it was transferred to Worcester, and later Gloucester, for the remainder of its main line career. This engine was recorded at Plymouth on the 10.35 a.m. Paddington on Saturday, 30th July 1938.

'Hall' No. 5961 *Toynbee Hall* approaching the S & D bridge at Cole on the six-mile climb to Brewham summit with the 11.45 a.m. Weymouth to Paddington service (Train No. 315) c. 1938. No. 5961 entered traffic in July 1936 at Stafford Road, but was transferred to Westbury in May 1938. Although 'Halls' and 'Saints' were to be regularly found on the Saturday train, Weymouth 'Stars' were the most common in the late 1930s. It is believed that the engine was scheduled to return with the 6.0 p.m. Paddington to Weymouth train, which was largely formed by the main portion of coaches (including the Refreshment Car) from the 11.45 a.m. Weymouth.

In addition to the Paddington services, through trains ran daily from Weymouth to Bristol and to Wolverhampton (via Oxford) in the latter 1930s. On summer Saturdays, extra trains ran from Weymouth to Paddington, Bristol, Birmingham and Cardiff, often hauled by '43XX', '29XX' and '49XXs'. This picture shows 'Saint' No. 2942 *Fawley Court*, from Bath Road shed, heading the 11.55 a.m. Saturdays-only Weymouth to Bristol (Temple Meads) train at Cole c.1938, running as usual via Westbury and Bath.

Old Oak 'King' No. 6003 *King George IV* with the Saturdays-only 12.10 p.m. Paddington to Plymouth (Train No. 170), a heavy service with coaches for Newton and Plymouth, and periodically for Minehead and Ilfracombe too. Official records for the late 'thirties indicate the train arriving at Plymouth behind 'Kings' or 'Castles' with 7 or 8 coaches, travelling unassisted over the banks. Old Oak at this time had 14 'King' class engines, one of which was earmarked for a Wolverhampton duty, leaving the serviceable balance for West Country services. In addition, Laira and Newton had a further ten 'Kings' between

COGLOAD AND CREECH

Cogload Farm occupational bridge was located on the short cut-off section between Athelney and Cogload Jct., just a half-mile to the east of the flyover. This aqueduct-style of accommodation bridge mounted on two pillars spanned a lengthy cutting, and features in many of Hubback's photographs of the area. 'King' No. 6001 *King Edward VII* from Old Oak, heading west with the 11.0 a.m. Paddington to Penzance train, was one of seven regular trains to leave Paddington between 10.25 and 11.10 a.m. for the West on peak summer Saturdays. This Penzance service ran on weekdays with additional coaches for Kingsbridge and Newquay, and on Saturdays with vehicles for Truro.

'King' No. 6002 *King William IV* passing Cogload Farm with the Saturday 10.35 a.m. Paddington to Penzance, c. 1937. When required, this train called at Newton to attach a banker (4–4–0 specified), which worked through to North Road. Leaving Paddington five minutes after the 'Limited', it was due at Newton nine minutes after the 'flagship' for the banker. Records of the late 1930s show this service to have been worked by either a 'King' or 'Castle' to Plymouth, with assistance from 'Bulldogs' over the South Devon banks. No. 6002 was from Laira shed.

Cogload Farm was also served by a road leading from the direction of Durston station, off to the right, which crossed the main line by the brick-built bridge featured here in this picture of Old Oak-based No. 6025 *King Henry III* heading the Up 'Torbay' (Train No. 511), 11.25 a.m. Kingswear to Paddington. On summer Saturdays, the regular 'Torbay Express' set of coaches was utilised for the preceding 11.50 a.m. Torquay to Paddington train, whilst the 11.25 a.m. Kingswear was formed by a set that had worked down overnight as the 12.20 a.m. Paddington; this accounts for the rather nondescript appearance of the train, which included a clerestory Van Third on the front. Both trains ran non-stop from Torquay to Paddington.

A train of new stock is featured in this view of No. 6018 *King Henry VI* as it passed under Cogload Farm bridge with the down 'Torbay', 12.0 noon Paddington to Kingswear, c. 1939. Like its up counterpart, the train ran non-stop between Paddington and Torquay, and included a two-car dining set serving luncheons (Table d'hôte 3s 6d, as on the 'Limited' – sixpence more than on other dining trains). The engine was from Newton, which regularly worked the up and down trains alternatively with Old Oak locomotives throughout the week. The farm occupational bridge over the cutting can be seen through the right-hand arch of the bridge.

Hubback took this picture of Old Oak 'Castle' No. 5080 *Ogmore Castle* with the 2.20 p.m. Minehead to Paddington service (Train 523) from Cogload Farm bridge in the summer of 1939. The full sweep of the curve from the junction flyover can clearly be seen, as the new 'Castle' headed east with another 2 hours 40 minutes of running ahead of it to London. This train traditionally changed engines at Norton Fitzwarren for its non-stop run to Paddington, rather than calling at Taunton to do so. The engine was probably scheduled to work down earlier that morning with the 9.35 a.m. Paddington to Minehead train as far as Taunton.

The quadrupling of the section between Norton Fitzwarren and Cogload Jct., a distance of about seven miles, and the reconstruction of Taunton station were carried out between 1930 and 1932. At the eastern end, the works involved the construction of a flyover at Cogload Jct. to carry the Down Bristol line across the 1906 direct line from Castle Cary, with connections between the Bristol and Main lines further to the west. Construction work was in its final phase when this picture was taken of No. 6020 *King Henry IV* from Laira easing a down express from Paddington under the recently-erected flyover one evening. The steelwork was carried out by the Horseley Bridge & Engineering Co. (Tipton & London).

An up train via the Berks & Hants route passing under Cogload flyover one summer evening c. 1931 behind Old Oak No. 6025 *King Henry III*. A variety of 70ft stock is in evidence on the train, which, by its formation, was possibly the 12.20 p.m. (Saturdays only) Newquay to Paddington; this service regularly had 'King' haulage at the time. Cogload signal box, shown here in its original position, was bodily moved to its new site by a pair of cranes in 1931, 'after having been moved three times previously' according to the *Great Western Railway Magazine*.

Laira-based No. 6002 *King William IV* restarting a down express near Cogload after being stopped by signals on Saturday, 1st August 1931. The sleepers and trackbed in the foreground were part of the work in progress on the flyover line from Bristol, which replaced the double junction that had existed since the opening of the Castle Cary line. Only two tracks were in use at this time.

The streamlined No. 6014 *King Henry VII*, speeding west near Cogload box in 1939 with the 11.0 a.m. Paddington to Penzance. This engine received its 'streamlining' in early 1935, but this was removed piecemeal through the years, with only the wedge-shaped cab front remaining beyond the war. Due to the 'bullet-nose' smokebox, the reporting number frame was carried on the front footplate. The Bridgwater & Taunton Canal kept close contact with the railway along this stretch of the route, running just to the north.

A new 'Grange' class 4–6–0 No. 6868 *Penrhos Grange* passing Cogload Jct. box on the up Bristol line with the 11.10 a.m. Paignton to Swansea (High Street) during the summer of 1939. The use of LMS stock, some of it vintage, in the front half of this train, was seemingly quite common on peak Saturdays. The engine, which entered service in March 1939, was stationed at Newport (Ebbw Jct) shed. Cogload box is shown here in its new position, a quarter-mile to the west of its original location. Durston box was only just over a mile away, and its outer distant was located beneath the Cogload home signal for the up Bristol line.

On the up main, Newton 'King' No. 6023 *King Edward II*, with the split bogie beam, passing Cogload box with the 11.50 a.m. (Saturdays only) Torquay to Paddington (via Lavington), c. 1938. This train was scheduled to convey six extra Thirds at the front of the train in addition to the regular eight-coach 'Torbay' set.

Laira-based No. 6004 *King George III* running past Charlton, between Cogload and Creech St Michael, with Train No. 146, the 11.0 a.m. Paddington to Penzance, c. 1936. This engine spent virtually the whole of its Great Western career running between Plymouth and Paddington, mostly via Lavington, but occasionally by the Bristol line. No. 6004 did spend three months working out of Exeter at the end of 1931, though it was still much involved with Plymouth trains. The train of LMS stock in the background was on the up relief/Bristol line.

The 1.30 p. m. Paddington to Penzance approaching Creech St Michael behind 'Castle' No. 4085 *Berkeley Castle*, c. 1936. At that time, the 1.30 p.m. was scheduled to convey a front section for Penzance (5 vehicles) then Newquay (1), Kingswear (2) and Ilfracombe (2), though this picture shows a further 4 unboarded coaches at the front. In July 1934, No. 4085 was transferred from Carmarthen (via Swindon Works) to Exeter, and on to Old Oak in May 1937 for an 18-month period.

No. 4090 *Dorchester Castle* at Charlton on the 10.35 a.m. Paddington to Penzance, c. 1936, with Cogload Jct. just visible in the distance. The engine was transferred from Penzance to Old Oak in December 1935 (via Swindon Works), and to Laira in April 1937, from where it carried out much work in Cornwall. Although by the spring of 1929 No. 4090 was paired with a 4,000-gallon tender, it ran with a 3,500 gallon one from the end of 1929 until April 1930 while working from Old Oak.

No. 6000 *King George V* running eastwards between Creech St Michael and Charlton at the head of the 11.0 a.m. Ilfracombe to Paddington (Train No. 516), c. 1937. This service was part of a divided Saturdays-only 12.15 p.m. Minehead to Paddington, which portion normally preceded it by some ten minutes on those occasions. The use of Old Oak 'Kings' on Ilfracombe trains from Taunton was established in the 1930 rosters (and possibly earlier), with the engine hauling a Plymouth to Paddington express to Exeter, and a local service to Taunton.

'Saint' No. 2942 *Fawley Court* (from Swindon shed) with the 10.20 a.m. Kingswear to Birmingham (Snow Hill), Train 555, on the up main approaching Charlton c. 1936. Most West Country & Midlands trains travelled via Bristol and Cheltenham, but this service was one of the very few that ran via Westbury and Oxford, though it did not serve any stations on the Northern line other than Birmingham. Bath Road 'Star' No. 4035 *Queen Charlotte*, alongside on the up Bristol line, was hauling the 10.30 a.m. Paignton to Liverpool, Train No. 580. The 10.30 Paignton had called at Taunton, and was switched to the relief line after, whilst the 10.20 Kingswear ran through non-stop on the main, although it was scheduled to be five minutes behind the Liverpool train at this point.

Old Oak-based No. 6029 *King Edward VIII* heading the 9.0 a.m. Perranporth to Paddington towards Charlton on the up main, c. 1937. The last station call of this train was Brent, where the scheduled Truro 4–6–0, which had worked through, was believed to have been changed for the 'King'. Sometimes, eastbound trains not stopping at North Road station changed locomotives at Laira Jct., though in this instance it may have been felt more efficient to change at Brent, where the two Kingsbridge vehicles (Van Third, Compo) were also attached to the front of the train.

Taunton was the gateway to the West Country, and on summer Saturdays witnessed a long procession of passenger trains destined for or originating from resorts in Somerset, Devon and Cornwall. The Victorian station, with its two through platform faces, was replaced in 1930–32 by one containing a central island platform between the two sides, giving four through passenger roads, again with bays on both sides at each end. This 1932 view of the west end of the station shows a gleaming 'King' approaching under the impressive semaphore gantry with the up 'Limited', 10.0 a.m. Penzance (due through Taunton at 2.20 p.m.). The train, which appears to have been formed with 1929 'Riviera' stock, was scheduled to convey through Brake Composites from North Road, Falmouth and St. Ives at the front, with the same stock Penzance portion at the rear. Taunton Station West box, built in 1931, is seen to the right and the new West Yard (with the goods running lines to the near side) on the left.

PASSENGERS
ARE NOT ALLOWED TO
CROSS THE RAILWAY
EXCEPT BY THE
SUBWAY

TAUNTON

In the early 1930s a 'King' was frequently used on the 10.40 a.m. Wolverhampton to Penzance onwards from Bristol; this train is pictured departing from the Down Relief line platform behind a 'King' class engine, possibly No. 6025 *King Henry III*, with the gantry signals clear for the Down Main, onto which the 'King' would shortly cross. Up to the summer 1930 timetable, there was a daily 9.45 a.m. Birmingham to Paignton, Penzance and Newquay train as well as the 10.40 a.m. Wolverhampton, both running via Stratford, but the former was subsequently altered to run from Wolverhampton (9.5 a.m.) on Saturdays only. The coach roof label board indicates 'Birmingham', which was the standard lettering until the later addition of 'Wolverhampton'.

'Star' No. 4014 *Knight of the Bath* from Shrewsbury shed standing in the Down Main platform at the head of Train No. 235, the 10.20 a.m. Liverpool to Paignton, c. 1938. No. 4014 was attached to a new 4,000-gallon tender (No. 2714) in May 1937, during her visit to Swindon factory, but reverted to a 3,500-gallon one in February 1939. The engine was transferred to Stafford Road shed in April 1939, being stopped there in April 1946, condemned at Swindon in June, and cut up in August of that year.

A '51XX' standing on No. 1 Carriage Siding with stock for a local train on Monday, 28th March 1932. At this time, '51XX' class 2–6–2Ts Nos. 5127 and 5132 were at Taunton shed, though with its curved drop end, this engine could possibly have been Yeovil's '5101' class No. 5169. The 'B-set' (two non-corridor Brake Compos) after the two 6-wheel 'Siphons', formed the basis of the services on the Barnstaple, Minehead, Yeovil and Chard branches out of Taunton. The lattice bridge in the distance carried the Minehead road out of Taunton, with the new Taunton West Junction box on the up side of the lines in front of it.

Taken from the long footbridge to the west of Taunton station, looking east, on Monday, 28th March 1932, this picture shows an auto on the Down Relief, probably waiting to cross to the up side. Auto cars at this time ran morning trips to Wellington and to Bridgwater, but were otherwise engaged on the Castle Cary or Frome duties. 70ft trailer No. 79 is seen here with a 'Metro' class 2–4–0T, which were a familiar sight around the area until the very end of the Great Western. A pannier tank is seen to the right of the gantry on the down siding with stock for (or from) the Barnstaple or Minehead departure bays which were located beyond.

The 'Torquay Pullman' was running during the early years that Gerald Hubback worked at Taunton, and passed the bottom of the garden of the house where he lodged (on the right). This train operated on weekdays during the summers of 1929 and 1930, with a limited service during the intervening winter period. The up train is seen here passing the up carriage sidings to the west of Taunton station with an unidentified 'Castle' in charge. Having worked down as the 11.0 a.m. Paddington to Paignton that morning, the train was scheduled to return as the 4.30 p.m. Paignton, due into Paddington at 8.30 p.m., with a formation of Brake Car Third, Third Kitchen, Third Parlour, First Parlour, First Kitchen and Brake Car Third. In 1929, the train carried an additional Third Kitchen and Parlour pair between the First class stock and the Brake. A rake of Paddington stock may be seen in the carriage sidings behind, with the old Taunton No. 4 box on the extreme left of the photograph.

No. 6010 *King Charles I* heading west on the Down Main near Fairwater Yard, Taunton, with the 12.3 p.m. Paddington to Torquay train, c. 1937. Although just three minutes behind the main train from Paddington, the 12.3 was scheduled to be five behind at Reading, and nine behind by the time Taunton was passed. The Minehead road bridge can be seen in the background, under which passed the four main running lines, and the up/down goods loop (seen here on the extreme right).

Landore-based No. 5047 *Compton Castle* heading a much-strengthened 1.30 p.m. Paddington to Kingswear, c.1938, past Fairwater Yard, Taunton. Numerous records show the engine on prewar Welsh expresses at Paddington, though there are no examples of the engine in the West Country at this time. In April 1940, No. 5047 was transferred to Newton, and became a familiar sight thereafter on the main line to the West.

NORTON AND THE WEST

On the approach to Norton Fitzwarren, Newton-based No. 5064 *Tretower Castle* is seen here hauling the 12.55 p.m. Cardiff to Paignton (Train No. 820) in 1937. Entering traffic in July 1937, this engine was renamed *Bishop's Castle* in September of that year, moving on to Shrewsbury around the same time. There was a daily through service between South Wales and Torbay all the year round at this time, with one or two additional trains on summer Saturdays.

The long footbridge at Norton Fitzwarren offered a wonderful vantage point for photographing trains at speed. This perfectly framed shot of Laira 'King' No. 6004 *King George III* passing with the 12.10 p.m. Paddington to Plymouth, taken in the summer of 1937, shows the quality of Hubback's work at that time. On summer Saturdays at this time, some 36 down and 33 up regular expresses were scheduled to pass this point between 10.0 a.m. and 4 p.m., added to which were divisions, excursions, local and branch passenger trains. At peak times, trains would follow each other at 6-minute intervals in each direction.

At peak periods in the latter 'thirties, the 1.0 p.m. Paignton to Wolverhampton (Saturdays only) was divided into as many as four parts, starting from Torquay, Teignmouth and Taunton in addition to the main train from Paignton. A 'Hall' is seen here passing through Norton Fitzwarren on the Up Main with Train No. 561, the Torquay to Birmingham (Snow Hill) portion, c. 1937. Periodically, the main train was formed partly of LNER stock, which was due to be handed over at Banbury after its visit to Wolverhampton. The '43XX' on the Up Relief, was waiting with what was probably the 12.50 p.m. Exeter to Taunton local, whilst in the yard, Old Oak 'Castle' No. 4073 *Caerphilly Castle* was awaiting the arrival of the 2.20 p.m. Minehead, which it would work on to Paddington.

Taken from the public footbridge to the east of the station, this photograph shows No. 5053 *Bishop's Castle* of Shrewsbury shed with the 9.10 a.m. Manchester to Paignton, conveying several 'vintage' LMS vehicles at the front of the formation. Entering traffic in June 1936, the engine was renamed *Earl Cairns* in August 1937, her original name being transferred to No. 5064. The '2301' class 0–6–0 and '45XX' 2–6–2T in the background had just drawn forward off the 2.20 p.m. Minehead, whilst *Caerphilly Castle* had drawn up behind them to set back onto their train.

No. 4073 *Caerphilly Castle* starting away from Norton on the Up Relief with the 2.20 p.m. Minehead to Paddington under clear signals. It would move across to the Up Main when clear of the public footbridge from which this photograph was taken. The engine had been changed at Norton because of the heavy occupation of Taunton station in mid-afternoon on summer Saturdays in the 1930s. Old Oak 4–6–0 engines were scheduled for many of the through Minehead and Ilfracombe trains, as were those from Bath Road, and occasionally Taunton. Surprisingly, records show that Landore engines also appeared, possibly being utilised by Old Oak shed prior to their return with a Welsh service.

The 10.40 a.m. Paddington to Falmouth is pictured to the west of Norton Fitzwarren, c. 1938, behind Newton-based No. 6023 *King Edward II*. In addition to the working of the 'Torbay', Newton's 'Kings' were also involved with Penzance trains to and from Plymouth. It was usual for a Laira 'Castle' to work the train onwards from North Road, though Exeter, Truro and Penzance engines (including 'Halls') were also recorded. The contrast in stock is interesting, with a clerestory coach and vehicles from the latter 1930s, including 'Centenary'. This train was effectively the fourth and last part of the down 'Limited', with services leaving Paddington at five-minute intervals from 10.25 a.m. on summer Saturdays. The bracket carrying the homes for the up line heralded the division of the running line into Relief (left) and Main (right) a short distance beyond.

To the west of Norton Fitzwarren, alongside the Barnstaple branch, No. 5919 *Worsley Hall* from Bath Road is shown heading the winter 1.15 p.m. Paddington to Penzance (via Bristol) express on Monday, 16th May 1932. This train was a fairly short-lived arrangement whereby the 1.15 p.m. Paddington (to Weston) and 1.30 p.m. (to Penzance, via Lavington) trains were combined, running via Bristol, with a slip portion for Bath, two vehicles for Weston, and one at the head for Ilfracombe, detached at Taunton. By Norton, the train was scheduled to convey a five-coach Penzance portion (including a dining car) and two for Kingswear at the rear, all 70ft.

The down 'Devonian' – 10.5 a.m. (Saturdays) Bradford to Paignton – running out of Wellington station behind Bath Road No. 2912 *Saint Ambrose* on Saturday, 13th July 1935. After a fast run between Leeds and Bristol (206 miles), the summer Saturday train was due out of Bristol (Stapleton Road) at 3.41 p.m. and into Paignton at 6.42 p.m. On weekdays, the train ran through to Kingswear, via Temple Meads. This service was scheduled for a curious formation in 1935, with a three-coach LMS vestibule/dining set sandwiched between GWR vehicles. After spells at Westbury and Swindon, No. 2912 was transferred to Weymouth in May 1943, remaining there until withdrawal early in 1951.

Laira-based No. 6016 *King Edward V* hauling the 1.30 p.m. Paddington to Penzance (Train No. 18s) away from Wellington on the same day. Ahead lay four miles of climb of around 1 in 80–90 to the summit beyond Whiteball tunnel, up which the 'King' was permitted to take 485 tons (around 15 or 16 coaches) unassisted. The Wellington down advanced starting signal, on the left, was placed on the far side of the line for sighting. An Old Oak or Laira driver working through would pass about 520 signals on the journey from Paddington to Plymouth.

One of the first Gerald Hubback photographs published in the *Railway Magazine* was this of Old Oak-based No. 6015 *King Richard III* climbing Wellington Bank to Whiteball Tunnel with a relief to the 11.10 a.m. Paddington to Penzance on Saturday, 13th July 1935. The train is seen passing the Whiteball down distant as it entered the lengthy cutting leading to the eastern portal of the tunnel. On a good day at this point, a 'King' could be travelling around 50 m.p.h. with the load carried, though it would generally lose a little more on the remainder of the climb.

No. 5022 *Wigmore Castle* from Old Oak shed with what may have been the 11.0 Paddington to Penzance on the final stages of the climb to Whiteball Tunnel, 13th July 1935. The gradient on the final westbound approach was 1 in 80, though it eased off to 1 in 127 at the tunnel. 'Castles' were permitted to haul 450 tons up the incline – 35 (one coach) less than the 'King' – to maintain the standard timings; the train here was probably not far from that limit.

Leaving the eastern portal of the 1,092-yard Whiteball Tunnel, No. 5004 *Llanstephan Castle* is seen descending the 1 in 80 gradient with a train of LMS stock on 13th July 1935. In both the preceding and following year, the train number 590 identified the 11.10 a.m. Paignton to Swansea service, which may well have been the case here. At peak periods, it was not unusual to see 'foreign' stock forming Great Western internal express services as a means of moving it towards the home system, minimising empty stock working. The engine was from Canton shed.

The Great Western introduced the system of train numbering for the summer timetable of 1934, initially for those services to and from the West Country only. On Saturday, 28th July 1934, Old Oak No. 6001 *King Edward VII* carried the number 150, which signified the 12.0 noon Paddington to Kingswear, the down 'Torbay', seen here leaving Whiteball Tunnel. The basic Saturday formation of the summer train comprised nine vehicles for Kingswear (including a three-car dining set) and two at the rear for Paignton, though this service had been strengthened by an additional two vehicles at the front on this occasion.

Probably the best known of Gerald Hubback's pictures is this view from the occupational bridge, looking through the Whiteball signals at 'Castle' No. 5028 *Llantilio Castle* with the 10.40 a.m. Paddington to Newquay (No. 130) on 28th July 1934. The train is seen passing the small signal box at the west end of Whiteball Tunnel, which marked the summit of the almost continuous climb from Norton Fitzwarren, though by far the steepest gradients occurred onwards from Wellington. Whiteball refuge siding, to the right of the train, was able to accommodate 44 wagons in addition to the engine and van, though with the Down Relief line a short distance further on, it was probably little used by through goods trains. Nevertheless, official reference is made to the holding of esparto grass or pulp trains there, awaiting acceptance at Silverton. The short engine siding can be seen to the left of the 'Castle'.

Looking west from the occupational bridge near Whiteball box, with Old Oak 'King' No. 6021 *King Richard II* in charge of the Up 'Torbay' 11.25 a.m. Kingswear to Paddington (No. 515) approaching Whiteball on 28th July 1934. The Down Relief line, on the left, was opened to traffic in 1927, and ran to the eastern end of Burlescombe station, some half-mile distant. During the 1930s, the 'Torbay' was normally worked by Old Oak and Newton engines alternately, though Laira would occasionally provide the power.

Running through the shadows of its own exhaust, Old Oak No. 6015 *King Richard III* is seen here pulling the 9.0 a.m. Perranporth to Paddington (No. 605) up the gradient on the approach to Whiteball box and tunnel on 28th July 1934; this was the third Saturday that the train numbering system had been in operation. On summer Saturdays, many of the West Country trains had Old Oak engines working eastwards at the end of the week's roster.

Slightly further to the west, Newton No. 6023 *King Edward II* is pictured working hard, though with steam to spare, as it passed under the bridge near Eastbrook with a portion of the up 'Limited' on Saturday, 28th July 1934. On the two following Saturdays, this train was worked by Old Oak 'Kings'. Eastbound trains were confronted with a 1 in 115 gradient for about two miles − rather less challenging than that westbound − on which the larger 4−6−0s were cleared for their normal maximum loads (530 tons in the case of the 'Kings'). The Relief line could house two trains under signals, the leading train of a maximum of 97 wagons and the following (nearer the tunnel) of 43, separated by the bridge featured here.

The complete train is featured in this study of a 13-coach down Sunday excursion, headed by No. 6010 *King Charles I*, as it approached Cowley Bridge Junction on 4th August 1935. The locomotive was from Laira shed, where it spent nearly the whole of its working life, being transferred to Old Oak in March 1959; in that year, most of the remaining West Country 'Kings' were moved to Old Oak or Stafford Road, although three remained into 1960. Much of the stock used for summer specials and excursions was stored during the winter months at various carriage depots and storage sidings throughout the system.

This final picture from the Gerald Hubback collection was taken during the early 1950s, and once again shows his partiality to the use of signals for visual effect. The up 'Torbay', hauled by Newton Abbot No. 4077 *Chepstow Castle*, is seen here on the Up Main near Obridge, framed by Taunton East Junction's down bracket signals in the final years of his lineside activity. The engine was at Newton during the early BR era until September 1954, but returned there from Laira in 1959.